AF270699

BIOLOGY PROJECT YOUR WAY

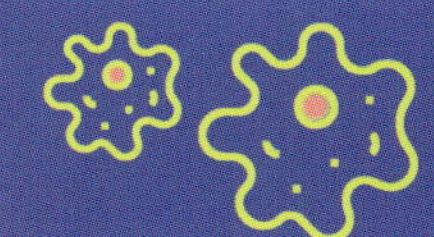

Megan Borgert-Spaniol

Super Sandcastle

An Imprint of Abdo Publishing
abdobooks.com

abdobooks.com

Published by Abdo Publishing, a division of ABDO, PO Box 398166, Minneapolis, Minnesota 55439.
Copyright © 2024 by Abdo Consulting Group, Inc. International copyrights reserved in all countries.
No part of this book may be reproduced in any form without written permission from the publisher.
Super SandCastle™ is a trademark and logo of Abdo Publishing.

Printed in the United States of America, North Mankato, Minnesota

102023
012024

THIS BOOK CONTAINS RECYCLED MATERIALS

Design: Aruna Rangarajan, Mighty Media, Inc.
Production: Mighty Media, Inc.
Editor: Liz Salzmann
Cover Photographs: Mighty Media, Inc.; Shutterstock
Interior Photographs: Adobe Stock, pp. 4 (bottom), 5, 28 (window); iStockphoto, pp. 4 (top),
6, 9 (orange), 10, 11 (left top, middle, bottom), 14 (bread), 15, 24 (both), 28 (left mold), 31;
Mighty Media, Inc., pp. 14 (soap, gloves, bags, bottle, box), 16 (experiment), 17 (experiment), 18,
21, 28 (experiment), 29 (experiment); Shutterstock, pp. 1 (microscope), 7, 8, 9 (girl), 11 (girl), 13,
16 (hand), 19, 22, 25, 26, 27, 28 (right mold), 29 (girl), 30
Design Elements: Shutterstock

Library of Congress Control Number: 2023939449

Publisher's Cataloging-in-Publication Data
Names: Borgert-Spaniol, Megan, author.
Title: Biology project your way / by Megan Borgert-Spaniol
Description: Minneapolis, Minnesota : Abdo Publishing, 2024 | Series: DIY science fair fun! | Includes
online resources and index.
Identifiers: ISBN 9781098292034 (lib. bdg.) | ISBN 9781098278939 (ebook)
Subjects: LCSH: Do-it-yourself work--Juvenile literature. | Biology--Juvenile literature. | Life sciences
--Juvenile literature. | Science projects--Juvenile literature. | Science fair projects--Juvenile literature.
Classification: DDC 507.8--dc23

Super SandCastle™ books are created by a team of professional educators, reading specialists, and content developers
around five essential components—phonemic awareness, phonics, vocabulary, text comprehension, and fluency—to assist
young readers as they develop reading skills and strategies and increase their general knowledge. All books are written,
reviewed, and leveled for guided reading, early reading intervention, and Accelerated Reader™ programs for use in shared,
guided, and independent reading and writing activities to support a balanced approach to literacy instruction.

CONTENTS

EXPLORE BIOLOGY

Do you love to watch birds or insects? Do you wonder how many **microorganisms** live in your garden? You might enjoy biology! Biology is the study of living things. Scientists who study biology are called biologists.

Biologists work on medicines to keep people healthy.

Biologists use electronic devices to track sharks and other animals.

Biologists **research** plants, animals, and microorganisms. They study what these life-forms need to live. They may try to find out how different life-forms affect one another.

Biologists use microscopes to observe bacteria and other microorganisms.

BECOME A SCIENTIST!

Scientists use a process called the scientific method. Check out the steps on the next page. You will use this method to **design** your own biology project!

THE SCIENTIFIC METHOD

1 ASK A QUESTION
What would you like to find out?

2 GATHER INFORMATION
What information do you need to understand your topic?

3 FORM A HYPOTHESIS
What do you think is the answer to your question?

4 EXPERIMENT
How can you test your hypothesis to find out if it is correct?

5 RECORD THE RESULTS
What did you observe in your experiment?

6 WRITE A CONCLUSION
Did your results support your hypothesis?

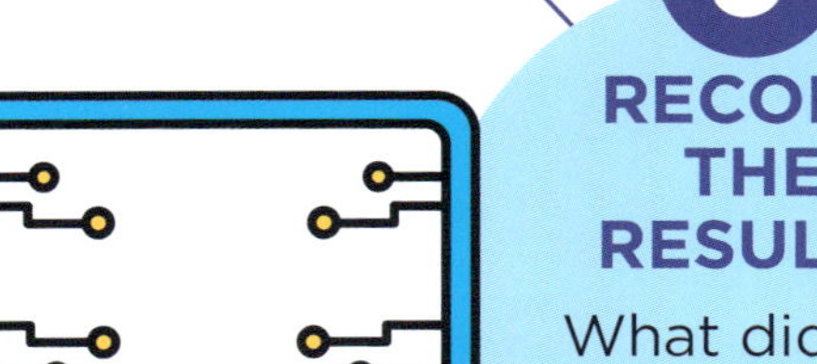

ASK A QUESTION

What topic do you want to learn about?

Maybe you are interested in how ants work together. Or maybe you want to know more about **microorganisms**. Start asking questions! Write your questions in a notebook so you don't forget them.

What kind of microorganism is mold?
What causes mold to grow?
9

GATHER INFORMATION

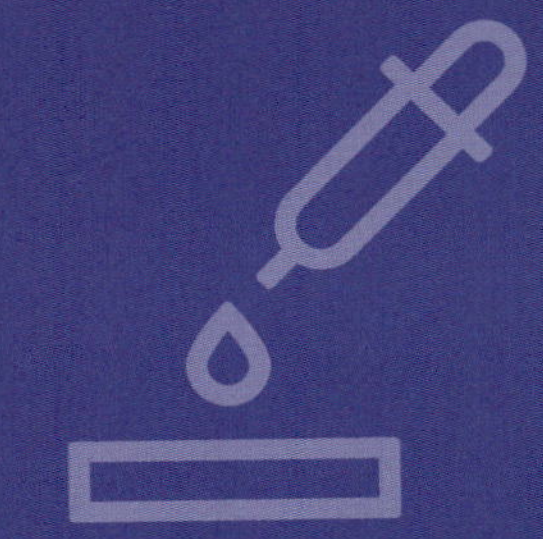

Scientists often have many questions they'd like to answer. But for now, choose one to **focus** on. Save the others for **future** projects.

It's time to **research** your **topic**. You can gather **information** from many different sources.

- **Read online articles about the topic.**
- **Read books about the topic.**
- **Talk to scientists or other experts.**

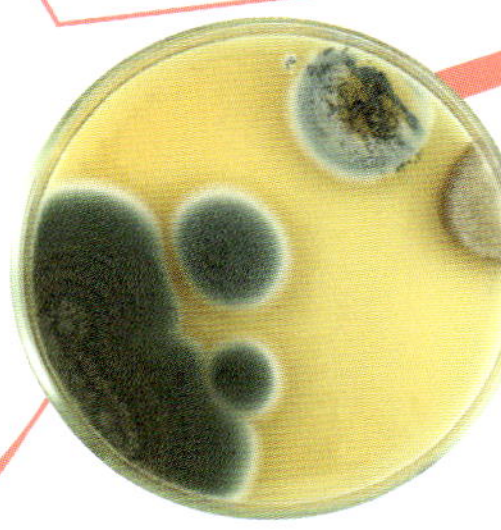

What did you learn about your **topic**? Write it down in your notebook. Then you'll have all the **information** you need in one place.

FORM A HYPOTHESIS

After you research your topic, it's time to form a hypothesis.

Your hypothesis is what you believe is the answer to your question. First, revisit your question. Do you want to change it based on what you learned? Then think of a few different hypotheses. Record them all in your notebook.

QUESTION: Does moisture affect the rate of mold growth?

HYPOTHESIS 1

Wet foods will mold faster than dry foods.

HYPOTHESIS 2

Dry foods will mold faster than wet foods.

HYPOTHESIS 3

Moisture will not affect the rate of mold growth.

Now, choose which hypothesis makes the most sense based on your **research**.
I learned that mold needs moisture to grow.
So, I think wet foods will mold faster than dry foods.
13

PREPARE YOUR LAB 

Find an area with a sturdy table or counter to work on. Then gather the supplies you'll need for your science experiment.

SUPPLIES

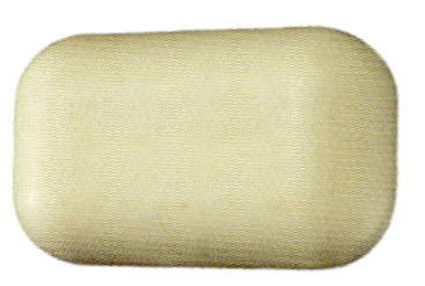

soap

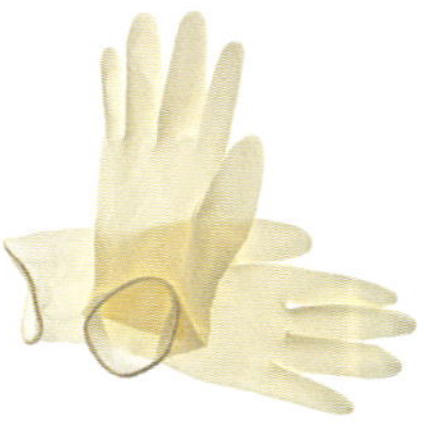

disposable gloves

bread that doesn't
contain preservatives

2 plastic bags

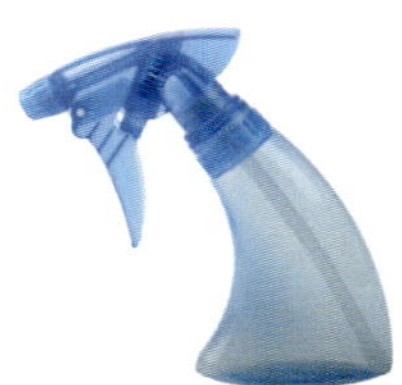

spray bottle
with water in it

cardboard box

All labs have rules that scientists have to follow. Here are some rules for your lab. They will help you stay safe and have fun while doing your experiment!

➡ **Ask an adult** for permission to use the materials and do the experiment.

➡ **Ask for help** with sharp or hot tools.

➡ **Wear goggles** and gloves to protect your eyes and hands.

➡ **Clean up** when you are done and put everything away.

EXPERIMENT!

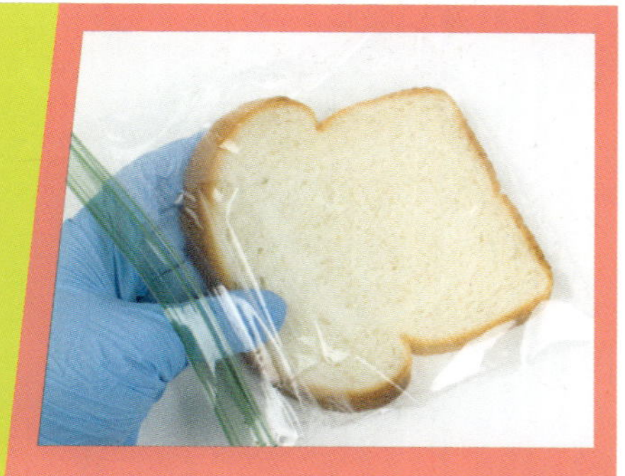

1 Wash your hands with soap and water. Then put on the disposable gloves. Seal one slice of bread in a plastic bag. Label the bag D for dry.

2 Put another slice of bread in the other plastic bag. Spray both sides of the bread with water. The whole slice should be wet. Seal the bag. Label it W for wet.

3 Put the slices of bread side by side in the cardboard box. Keep the box in a warm room.

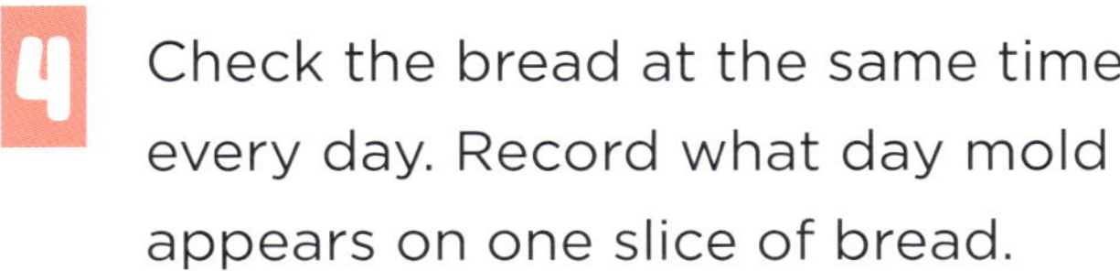

4 Check the bread at the same time every day. Record what day mold appears on one slice of bread.

5 Record what day mold appears on the second slice.

Look at the results of your experiment so far. You might be ready to draw a conclusion. But first, consider any **variables** that might affect your results.

Touching food with bare hands can contaminate it with **microorganisms** that cause mold growth.

You washed your hands and wore gloves to avoid contaminating the bread.

Both slices of bread were exposed to the same temperature and amount of light.

RECORD THE RESULTS

During experiments, scientists record data and other observations. You wrote down how long it took for mold to grow on each bread slice. Now it's time to record your data to share with others.

Scientists often use tables and graphs. This helps make the results easy for others to understand.

A table organizes **information** in rows and columns.

BREAD SLICE	DAYS BEFORE MOLD APPEARS
Wet	4
Dry	6

A bar graph compares data using rectangular bars.

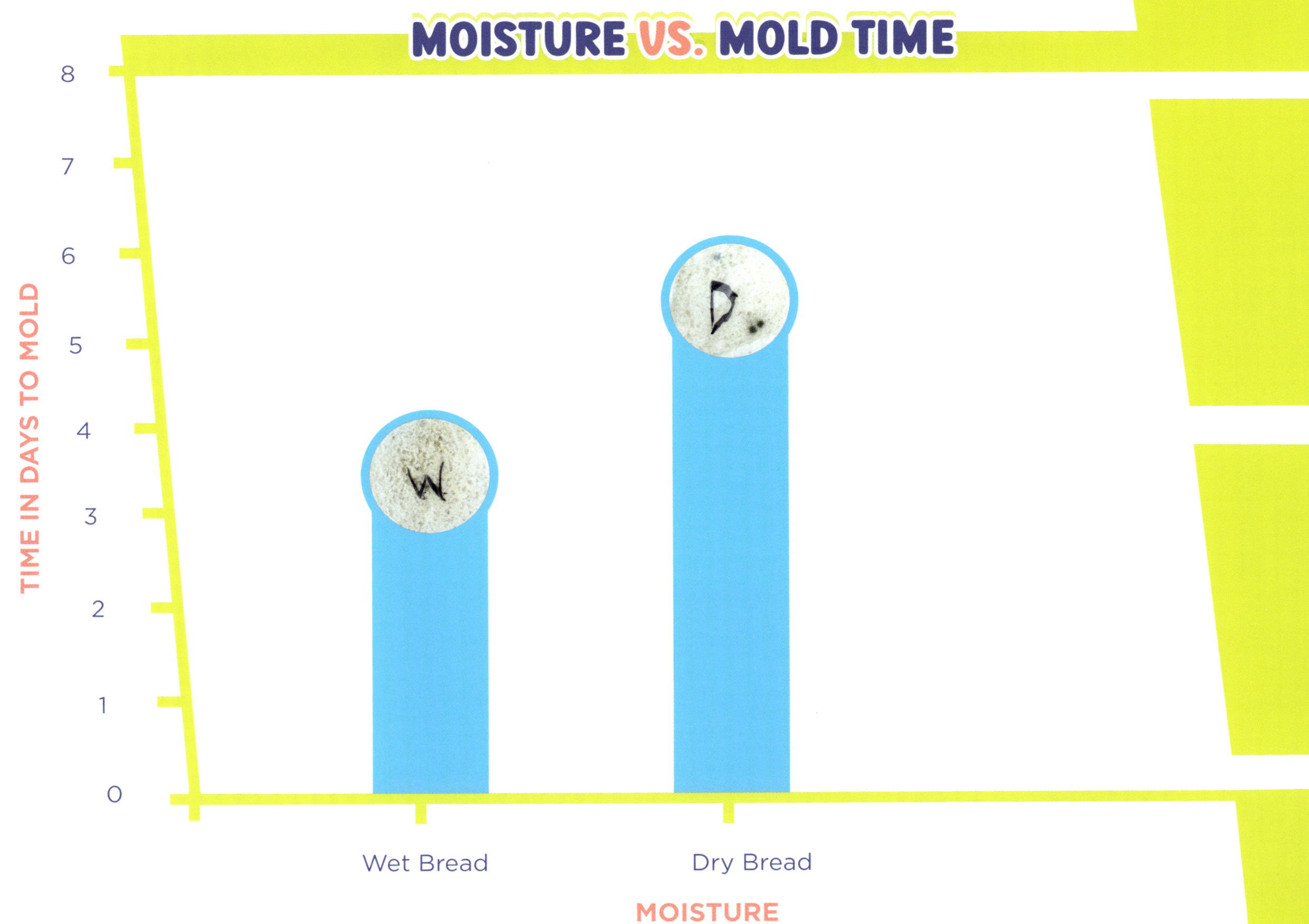

WRITE A CONCLUSION

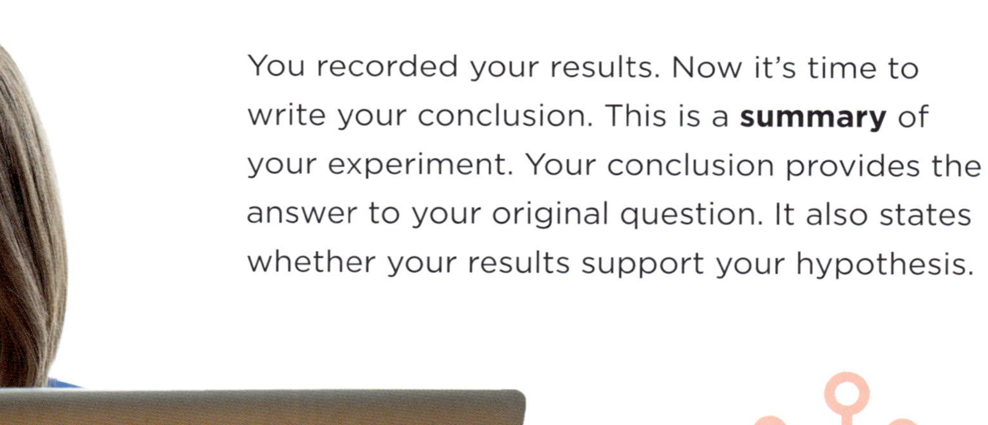

You recorded your results. Now it's time to write your conclusion. This is a **summary** of your experiment. Your conclusion provides the answer to your original question. It also states whether your results support your hypothesis.

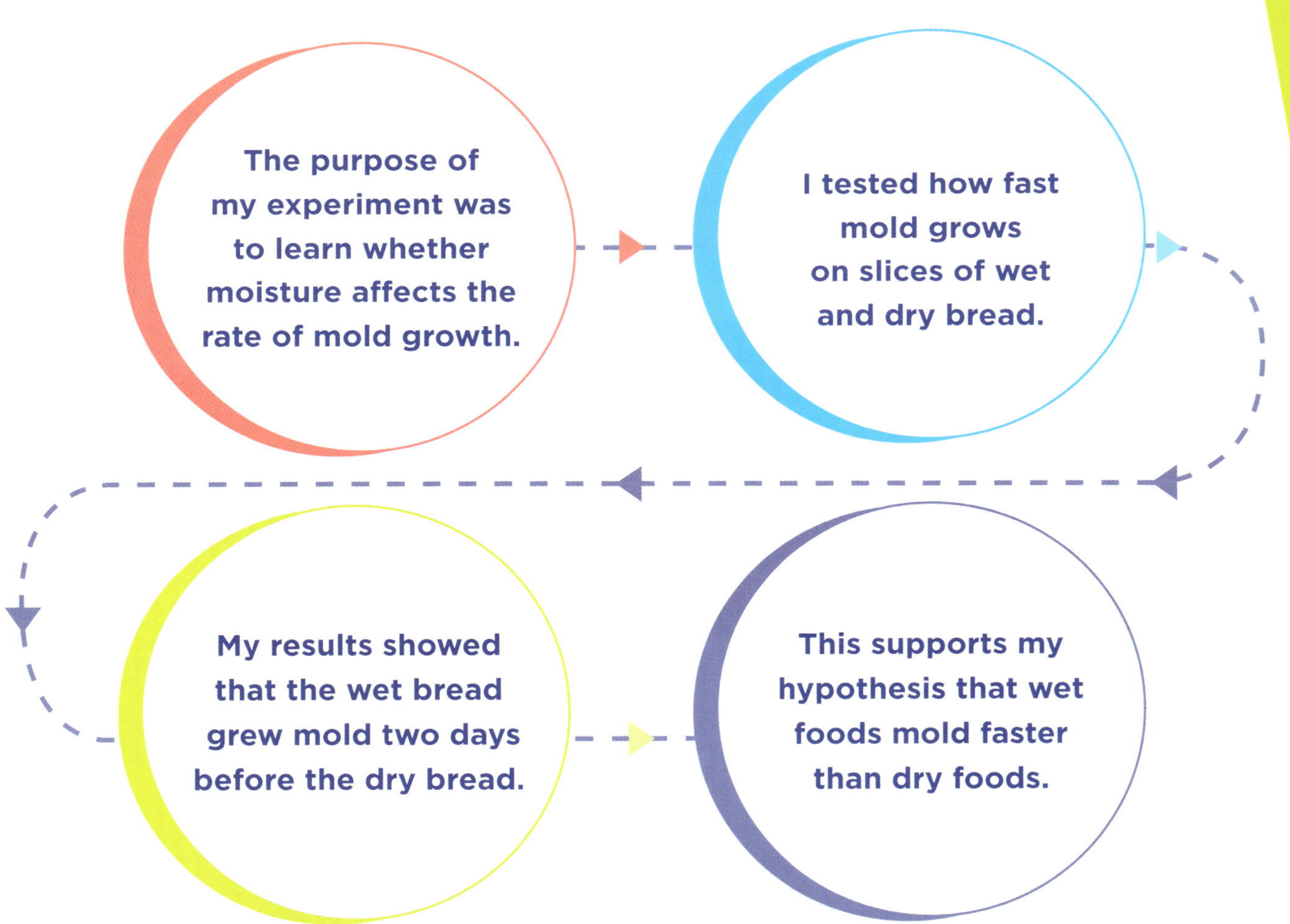

Many scientists find that their hypotheses were wrong. There is nothing wrong with being wrong! It means you did your experiment without **bias** and were surprised by the results. That is another mark of a great scientist!

FURTHER RESEARCH

A conclusion offers an answer to your original question. But it can bring up new questions too! For scientists, the end of one experiment often leads to more **research** and new hypotheses to be tested.

What new questions do you have? What could you research further? Do you have a new hypothesis to test?

Does temperature affect how fast mold grows?

Does the amount of salt in a food affect how fast mold grows on it?

What conditions lead to the most mold growth?

PRESENT YOUR PROJECT

Young scientists share their **research** at science fairs. Students share what they learned with classmates, teachers, parents, and sometimes judges. It is a chance to show all the work they put into their experiments.

Demonstrate or show a video of your experiment.

Create comics or other drawings to show your project in a fun way.

Include props, models, or dioramas.

One way to present a project is with a display board. It should show how you followed the scientific method. Turn the page to see a display board of the project in this book!

QUESTION

Does moisture affect the rate of mold growth?

RESEARCH

Mold is a type of fungus. Fungus is a type of microorganism. Mold is made up of tiny spores. Large groups of spores can be seen without a microscope. Mold needs moisture to grow.

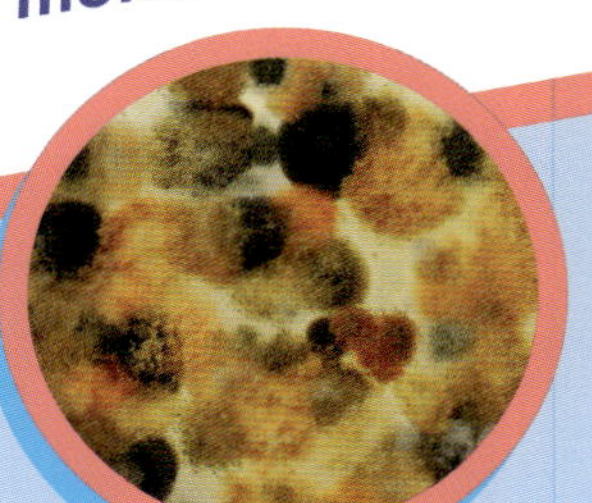

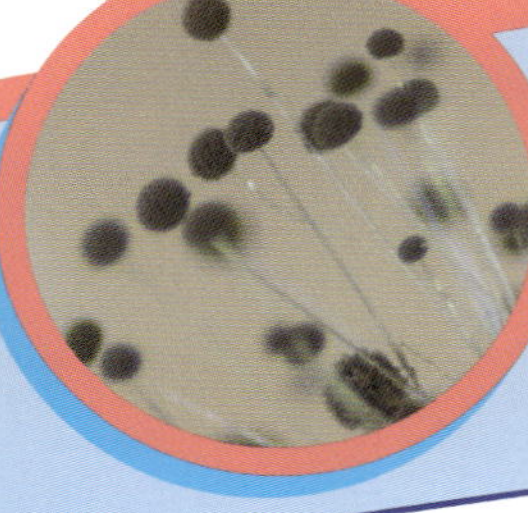

HYPOTHESIS

I think wet foods will mold faster than dry foods.

EXPERIMENT

I tested how fast mold grows on slices of wet and dry bread. I placed the slices in separate bags. Then I recorded when mold started to grow on each one.

RESULTS

My results showed that the wet slice of bread grew mold two days before the dry slice.

CONCLUSION

The results support my hypothesis that wet foods mold faster than dry foods.

KEEP ASKING QUESTIONS

Your science project is over. You packed away your display. But don't stop asking questions! What might you do differently if you did the project again? What additional **research** could you do? Is there a related **topic** you would like to explore?

Beyond the Science Fair

Be a scientist beyond the science fair! You can use parts of the scientific method to find answers to everyday questions. Maybe you have a hypothesis for why your plant is turning brown. Maybe you experiment to learn which seeds your pet bird likes best. One day, you might use science to do big things. Maybe you'll help cure diseases! Turn your world into a science fair. What will you discover?

GLOSSARY

bias—showing a preference for one result over another.

design—to plan how something will appear or work.

focus—to concentrate on or pay particular attention to.

future—the time that hasn't happened yet.

information—the facts known about an event or subject.

microorganism—a living thing that is too small to see with the naked eye.

research—to find out more about something. Also, a study of something to learn new information.

spore—a small, single-celled organism made by plants such as fungi that can grow into a new plant or organism.

summary—a short statement of the main points.

topic—the main idea or subject.

variable—a factor in a scientific experiment that may change.